Should I Take the Shot?

Practical and Ethical Considerations for Shooting a Bigfoot

ISBN: 1-4392-0954-5

To order additional copies, please contact us.

Booksurge Publishing
www.booksurge.com
1-866-308-6235
orders@booksurge.com

TABLE OF CONTENTS

INTRODUCTION..1

    Description of the Bigfoot...................................1

    The Elusive Bigfoot...2

    The Killing Controversy....................................3

WANTED DEAD OR ALIVE: A BIGFOOT SPECIMEN..........5

    The Value of a Dead Specimen..........................5

    The Value of a Live Specimen...........................9

BIGFOOT HUNTING..11

    Warning...11

    Taking the Creature Alive................................11

    Taking the Creature Dead................................14

    Preliminary Considerations..............................14

SUMMARY..23

CONCLUSION...27

REFERENCES..29

INTRODUCTION

Description of the Bigfoot

The animal known as Bigfoot (a.k.a. Sasquatch) has been reported as a large seven- to nine-foot tall ape-like beast that weighs between three hundred and nine hundred pounds. Bigfoot is reported to have reddish-brown or auburn-colored hair covering the majority of its body. The creature is described as having broad shoulders, long and thick arms, a nonexistent neck, flat face, sloped forehead, brow ridges, a cone-shaped head, and a horrible odor. Sightings of both sexes of the Bigfoot animal and their offspring have been reported. These creatures are generally reported to be solitary in nature and are rarely seen in the company of others. Communication among the species is reported to consist of a series of grunts, howls, shrieks, or whistles.

Footprints of the Bigfoot animal range between eleven to twenty-one inches, with the average footprint being sixteen inches long and seven inches wide. The distinguishing characteristics of these footprints include three, four, or five toes, no foot arch, and a pronounced double-ball heel. It is assumed that the variations that occur in the creatures' physical description, as well as that of the footprints left behind, are because several different species of the creature are believed to exist.

The Elusive Bigfoot

A common question that often arises about Bigfoot is, "If a Bigfoot exists, why hasn't anybody killed one yet?" An obvious response to this question may be shock. A Bigfoot is an unfamiliar creature, and whenever encountering an unfamiliar situation, our mind goes into a state of shock. For example, I was once driving down a country road, and from a distance I saw a deer in the middle of the road. This contrasting image of a wild deer standing on a modern road so totally befuddled my senses that even though my eyes saw the deer, my rational mind did not wish to accept the incident as even being possible (thinking perhaps that it was a hallucination). Finally, I slowed down and stopped my car as the deer trotted away. This could be the same reaction that occurs when a human encounters a Bigfoot. As the person is stunned by the extraordinary, the Bigfoot, who is more familiar with the appearance of man, makes a hasty escape.

Therefore, should a human and a Bigfoot ever surprise one another, I believe the Bigfoot will always have the first-move advantage (a serious drawback for the human) to either fight or flee. As the Bigfoot is generally reported to flee immediately from encounters, this reason probably explains why the Bigfoot is so elusive.

The Killing Controversy

The question of 'to shoot or not to shoot a Bigfoot creature should you encounter one' has been a heated debate for many years. For years Bigfoot sightings have been claimed and reported, but the existence of this elusive creature has yet to be verified.

Arguments for shooting the Bigfoot include the belief that once a carcass of a Bigfoot is brought before the feet of the scientific community, the existence of the animal can then be unquestionably confirmed. Then the corpse could be studied to determine why the creature has become endangered and what preventative measures could be taken to ensure that the animal does not go into extinction.

Arguments against shooting the Bigfoot creature cover many areas of beliefs and traditions. Firstly, the hunter is aware that shooting a Bigfoot violates poaching regulations and may incur a stiff penalty. Also, as the Bigfoot creature has been reported as being excessively shy around humans by generally fleeing from them, the senseless slaughter of such a timid creature may be viewed as a most heinous crime against nature. Additionally, the person who brings to the world irrefutable evidence of the existence of Bigfoot may become despised by tourists, 'tourist-trap' business owners, hunters, historians, researchers, authors, tall-tale tellers, and others who wish to keep the Bigfoot fable alive. These and other reasons not to shoot a Bigfoot creature are discussed in further detail in this work.

WANTED DEAD OR ALIVE: A BIGFOOT SPECIMEN

The Value of a Dead Specimen

Even if a Bigfoot corpse were brought before the scientific community for examination, would sufficient proof of the creatures' existence be validated? Skeptics among the scientific community may still reject the dead body as an actual Bigfoot and accept it only as an ape with a unique genetic birth defect, thereby still keeping the Bigfoot legend alive. So in order to prove that this is not a singular anomaly, several additional specimens would need to be brought in for comparison. This situation could lead to genocide of the Bigfoot species With physical evidence being a possibility, hordes of hunters motivated by fame and fortune would load their rifles and search for the next Bigfoot to kill. With such a large reward to be won for the next Bigfoot corpse brought in, ruthless competition among hunters would surely lead to arguments as to who's bullet actually inflicted the killing shot.

However, even if the corpse of a true Bigfoot was brought before the scientific community, the mystery behind its endangerment may still remain unsolved. Although we think of the creature as being related to humans or apes based solely on reported sightings, its origin, species, diet, and other characteristics are yet unconfirmed. Perhaps the internal organs of the creature are completely unrelated to any other man or animal previously examined. Also, we could only associate the current medical condition of the dead creature to what we already know about similar creatures. For example, if a human cadaver is examined and found to have clogged arteries, we can say that individual had a diet high in cholesterol due to the consumption of red meat. By applying similar logic, if a Bigfoot cadaver is also found to have clogged arteries, we would naturally make the assumption that Bigfoot is a carnivore also. On the contrary, Bigfoot could actually be an herbivore (plant eater) but due to its unique body chemistry could produce elevated cholesterol levels from what would normally be considered healthy eating habits according to human and primate body chemistries. Furthermore, due to the creatures' similarities to humans and apes, scientific examinations of these creatures would draw too heavily on previous research, and conclusions would be made without performing the necessary validations. For example, consider this falsehood in logical reasoning: if body activity A occurs due to the existence of amino acid B and amino acid C in the human body, then by accepting the fact that Bigfoot is related to humans, if body activity A occurs in Bigfoot and amino acid B exists within his bloodstream, then amino acid C must also exist in Bigfoot's bloodstream. As should be apparent, jumping to these conclusions could be a mistake when dealing with the unknown. Perhaps due to Bigfoot's unique body chemistry, its body could perform bodily activity A with an amino acid other than C or perform the activity using amino acid B alone. Furthermore, the activity may not even depend on either amino acid. We do not have to be medical specialists to see the logical paradox here and its predictable outcome. The medical community would be so eager to relate the creature to a human or an ape that they would attempt to draw conclusions based solely on what we know about these similar animals.

Likewise, the corpse of a dead Bigfoot may prove practically useless if its bodily organs and functions vary greatly from what we are already familiar with. For example, consider an alien autopsy. During the autopsy, the medical examiners would speculate on what the internal organs of this extraterrestrial creature are and what they do, but one can only assume that their associations are based upon what we already know of terrestrial creatures (for example, 'This orange square thing must be the alien's stomach.').

Furthermore, the carcass of the dead animal can only provide information about the physical characteristics of the creature, not its intellect, cognitive abilities, nor social structure. It could be possible that they are becoming extinct not from changes in their environment but of their own accord. It is possible that the Sasquatch do not possess the necessary survival intellect not to eat poisonous berries or snakes, or to stay out of inclement weather, or perhaps they possess poor social behavior among members outside of their groups and fight and kill among themselves. Even though the Bigfoot has been reported as being a gentle and timid creature, usually running away from humans during an encounter, it does not imply that they would not attack a member of their own species.

Another point to consider is that a single specimen of a creature does not necessarily represent the entire species. For example, suppose that a human who is in poor physical health or an abuser of alcohol or drugs landed on an alien planet. The occupants of the alien world would incorrectly accept this one specimen to represent the physical condition and behaviorisms of the whole human race.

The Value of a Live Specimen

Should a live Bigfoot ever be captured and placed within a habitat so that it could be studied, the actions of the captured Bigfoot would only reflect the characteristics of the individual specimen and not necessarily those of the entire species. With just one Bigfoot under observation, only information about that one subject and sex could be studied. No knowledge could be acquired about the social behavior of the captured Bigfoot.

Also, should the captured Bigfoot have the intellect to acknowledge that it is existing within an 'artificial world,' it may behave in a different manner, thereby behaving uncharacteristically before the observers. As humans, we relate to the world according to how we perceive it, and we try to simulate those conditions within a natural habitat for the creature under observation. However, humans may not perceive the environment at the same level as the animal under observation. For example, consider the modern canine, which can distinguish and even track a specific person through a highly acute sense of smell, even though we fail to recognize any discernible odor among ourselves. Similarly, even though tap water and natural spring water smell the same to humans, the Bigfoot may discern a difference when we 'open the hose' for the natural spring water in the habitat.

Finally, we must again consider if the captured specimen in this case represents a true example of the species. It is possible that the captured specimen could be suffering from any number of unique physical, mental, psychological, social, nervous, or other disorders. These disorders could have been inborn or manifested later in life due to difficulties that this singular Bigfoot experienced.

BIGFOOT HUNTING

Warning

When hunting a Bigfoot, not only the health of the creature but that of the hunter lies in harm's way. The Bigfoot has been described as a seven- to nine-foot tall man-like ape weighing around six hundred pounds with long, powerful arms. Although this creature has been reported as being timid and flees in the presence of man, one can only imagine the magnitude of destruction that such a beast is capable of if threatened or physically harmed. The Bigfoot creature can either be taken alive or dead.

Taking the Creature Alive

Generally, in order to obtain a living Bigfoot specimen, tranquilizer darts and injections would be utilized. Through the use of tranquilizers the beast may be either sedated so that blood and tissue samples can be safely obtained before the animal is set free, or the creature may be physically abducted. Should biological samples be extracted, there is no assurance that they would not become lost, stolen, or damaged in transit. Also, the creature may have an adverse reaction to the tranquilizer drug. In the end, the samples may not provide conclusive proof of the creature's existence.

If the intent is to abduct the creature, you could create an ambush by encircling the creature with riflemen ready to shoot tranquilizer darts until the creature is sedated. With the creature sedated, it should then be bound with chains and placed in the back of a large off-road vehicle and transported to civilization.

As the Bigfoot is of such immense proportions, the physical capture of such a creature by a single person would prove practically impossible. Therefore, a team of personnel should be employed to capture the Bigfoot. This team of personnel, however, would be hard to assemble because Bigfoot hunters are often scoffed at and ridiculed by the general population. The members of the team would probably just be 'in it for the ride' or the money and not take their responsibilities seriously. They may even go as far as damaging valuable equipment out of contempt for the hunters. Furthermore, the riflemen that are hired may prove themselves unreliable. Instead of shooting at a Bigfoot, they may merely drop their guns and run away in fear.

But even if you are successful in assembling a reliable team, all of your expenses and efforts might still go to waste as you may never spot a Bigfoot during your hunting expedition. The Bigfoot has been successful at avoiding being spotted by large crowds of people and has only been witnessed, photographed, and filmed by individuals or small groups of people. Perhaps the Bigfoot has an acute sense of smell and recognizes the odor of humans and thus keeps its distance from large groups of people. The odor of just a few people, however, may escape its detection.

In any case, tranquilizing a Bigfoot is a risky endeavor. As the Bigfoot is an unknown animal, the effectiveness of any tranquilizer at any dosage cannot be assured. Should you use the wrong type or too little of the tranquilizer, the beast may become enraged and attack the hunters. On the other hand, should the type or dosage of the tranquilizer be too strong, it may kill the animal.

In the event that the hunters successfully tranquilize the animal and wish to transport it to civilization, a constant watch on the creature must be maintained to ensure that it remains asleep and does not awaken in a violent rage. This action may eventually lead to an overdose of the tranquilizer drug and the death of the animal.

Taking the Creature Dead

Preliminary Considerations

Before committing yourself to taking a Bigfoot dead, there are several factors to consider. Firstly, consider the fact that some locations have laws that protect the creature from human harm. For instance, the law enforcement in Fouke, Arkansas, strongly discourages hunting the creature for reasons of safety and other concerns. Additionally, you may find yourself charged with poaching or violations of firearm regulations.

Apart from the legal violations, you may also find yourself to be in physical danger from the beast itself. Although most high powered rifles are capable of inflicting massive damage, they generally have a slow rate of fire and generally only hold a few shells per magazine. For example, a typical 0.308 rifle can deliver a powerful shot, but because it is a bolt-action rifle, the rate of fire is very slow, and the magazine only holds three shells. This may not be enough fire power to take down a three hundred- to nine hundred-pound wounded beast charging at you with a three- to four-foot stride. So armed with this particular rifle, you would opt for the 'one shot, one kill' approach and aim specifically for a vital target area. Realize, however, that although the location of vital organs of known species is readily identifiable, the internal organ structure of an unknown creature such as Bigfoot may be slightly or even significantly different than what we would normally expect (although a shot to the head should prove lethal to just about any animal). This kind of precise shot may become extremely difficult to place, particularly at close range before an enraged and rapidly approaching creature. Also remember, as stated previously, the Bigfoot would generally have first-move advantage.

On the other hand, if you were to go Bigfoot hunting with an M-60 machine gun, you could probably reduce the creature to shreds in moments. Although you would be thankful that the gun saved your life, the scientific community would not appreciate you bringing before them a bullet-ridden (and perhaps half-blown away) Bigfoot corpse, as they would only have fragments of the creature to examine.

Although the prize of fame and fortune would be awarded to the individual who shoots a Bigfoot creature and presents it to the scientific community for examination, these prizes do not come without their repercussions. Generally, we desire fame for the positive attention we get from others. When you are well known in society, strangers will know who you are, want to talk to you, and send praise in your direction.

Unfortunately, there are negative consequences to fame. You need to realize that in your climb to fame, you are directing the spotlight away from those who are already there. These famous people may throw dirty tricks, insults, and slander against you so that the spotlight returns onto them. They realize that if they do not keep the spotlight always focused on themselves, they would merely fade from the public eye. In an attempt to reestablish their own popularity, they will use whatever tactics necessary to sway a fickle public to switch their view of you from being a hero to that of being an inhumane murderer.

Also realize that due to the fierce competition that exists today, whatever attention you do receive will probably be extremely short lived. To regain your fame and popularity, you would always find yourself competing for the spotlight by 'doing the next big thing.' Are you willing, able, and ready to accept strong feelings of great admiration from some of society and bitter hatred from others? Do you wish to always be in competition to stay in the spotlight? Or are you just happy to receive a neutral response from others and stay out of the spotlight?

People will always associate fame with fortune and money, a commodity that everyone desires. Having lots of money usually means that a person can have the freedom to live the way they wish. With a great deal of money we can quit ours jobs, pursue our own interests and hobbies, buy the items that could enrich our lives, and travel to whatever destination we choose.

Unfortunately, there is a dark side to having lots of money. Suspecting that you have money, people would be all too happy to help you rid yourself of it through legal or illegal means. Scammers,

17

thieves, crooks, 'gold diggers,' and even kidnappers would all be alerted to your newfound wealth and will attempt to gain a piece of it. Thus extremely wealthy people often find it necessary to hire a bodyguard service for themselves and their families. Not only does this service interfere with their personal and social lives, but it can be quite expensive to maintain. Can you afford to maintain this bodyguard service after your fame fades away, or are you going to go on chasing the spotlight to continue your fame and fortune? Everybody is different. For some, fame and fortune serves them well. For others, fame and fortune may just bring different kinds of problems.

Before bagging yourself a Bigfoot, it would be a good idea to reflect upon society's possible negative opinions of your action. Bigfoot is generally not known to terrorize people or villages. If it did and you hunted down and killed the pesky creature, you would be hailed as a hero and a savior. Most believe, however, that Bigfoot has coexisted peacefully and inconspicuously with humanity throughout the centuries. In fact, reports of Bigfoot sightings describe the creature as a big and gentle animal with a shy personality who generally tends to run away from humans. Thus if anyone ever were to bring in a Bigfoot corpse, it would likely to have been shot in the back as it was fleeing. This action would be seen by many as a most heinous crime against nature and an act of cowardice. However, should ballistic evidence prove that all shots entered the body of the creature from the front, the hunter's claim that he or she shot the beast in an act of self-defense would be uncontested.

Other people may also tend to disfavor the person who brings in a true Bigfoot specimen for examination. Take the following, for example:

1.   Those individuals who like to believe that there are yet some undiscovered mysteries to our planet. These people despise the fact that everything has already been thoroughly examined and classified into 'neat little categories.'

2.   People who believe that their faith in the fact that the Bigfoot creature exists and don't need to see physical proof.

3.   Scientists and authors who have speculated on the characteristics and nature of the Bigfoot who may lose their credibility once a live specimen is examined.

4.   People who make a living by keeping the Bigfoot legend alive may lose their livelihood should a live specimen ever be produced.

5.   Bigfoot hunters who may now have to search for a different cryptid.

There are some additional questions to consider before killing a Bigfoot. Would you be killing the last remaining Bigfoot, the mate, or perhaps the parent of dependent children, thereby being responsible for driving the species into extinction? Could the extinction of the Bigfoot species lead to an unfortunate disruption in a balanced ecosystem? The final question you should consider is, 'Am I sure this Bigfoot is alone?' As Bigfoot creatures have been rumored to live within small social groups, be prepared to defend yourself should another Bigfoot animal come out of hiding seeking vengeance for its fallen companion.

Now you are aware of the dangers and repercussions that may

occur should you be successful in bagging a Bigfoot. If you are still determined in your course of action, here are some final suggestions.

To take the Bigfoot dead you would have to carry conventional firearms and ammo on your hunting expedition. First of all, you would need to be sure that your ammunition is of the proper caliber for the intended rifle and that it has sufficient power to take down the beast. You will also want to familiarize yourself with all of the wildlife within the hunting area so that you don't shoot a wild ape or other animal by mistake and be charged with poaching. Shooting any animal, including a Bigfoot, out of season may be considered poaching and could result in steep fines and penalties.

You should also be psychologically conditioned so that you do not go into shock at the sight of the monstrous creature. On the other hand, you do not want to shoot so readily that you shoot another animal or human by mistake. By becoming well acquainted with photographs and other images of a Bigfoot, you should not only easily recognize one when you meet one in the flesh, but you should be more psychologically prepared for the encounter so that you can minimize your shock-recovery time when first encountering the creature.

This shock reaction will have less effect as the distance between you and Bigfoot increases. As the distance increases, your perception of being in immediate danger by this formidable creature lessens, thereby reducing your recovery time and giving you more time to steady your nerves, focus your thoughts, aim your rifle, and take the shot. Therefore, during the hunt you should keep scanning the entire area in which you are traveling, scanning close by to avoid any surprises, as well as into the distance either to get an opportunity shot at a Bigfoot or to catch it as it flees.

After the beast has fallen, it would be wise to delay approaching it to ensure that it is in fact dead. Then to ensure that you did not merely knock it unconscious, you should restrain the beast with chains. Finally, load the creature onto a heavy-duty off-road vehicle and return to civilization for your fame and fortune.

Happy hunting!

SUMMARY

Sightings of a Bigfoot have occurred throughout the centuries. The beast has been primarily reported within Northern America and Canada, along with a few international sightings. Reports describe the creature as a large seven- to nine-foot tall ape-like beast weighing between three hundred and nine hundred pounds. Footprints of the Bigfoot average sixteen inches long and seven inches wide. The creature is said to have a shy and timid demeanor and flees at the sight of man.

The controversy over whether to shoot a Bigfoot creature should you encounter one has been a heated debate for many years. Although Bigfoot sightings have been reported throughout the centuries, no physical specimen of this elusive creature has ever been discovered.

Arguments for shooting the Bigfoot include the belief that once a carcass of a Bigfoot is brought before the scientific community, the existence of the animal could then be unquestionably confirmed. With a physical specimen for analysis, the scientific community could then hypothesize the cause of their endangerment and could suggest ways to prevent the extinction of the species.

Arguments against shooting a Bigfoot cover many aspects: safety, legality, beliefs, and traditions. Firstly, the creature may become dangerous if sufficiently provoked. Hunters may also incur legal penalties, including violations of poaching and firearm regulations, by shooting a Bigfoot. Additionally, the hunter who successfully kills a Bigfoot may face persecution for his or her action from humanitarian organizations, celebrities, the press, 'tourist-trap' businesses, authors, scientists, as well as other individuals who wish to keep the Bigfoot legend alive.

Unfortunately, even if a Bigfoot corpse were brought before the scientific community for examination, proof of the creature's existence may still be up for debate. The specimen may be either accepted as a true Bigfoot specimen and declared as the last of its kind or totally rejected as a Bigfoot and merely declared to be an ape with a unique genetic birth defect. Additional specimens would need to be collected for comparison analysis. Furthermore, inaccurate conclusions may be derived through comparison of this newly discovered species with what researchers may consider to be related species.

Although the behavior of a live Bigfoot could be studied within an artificial habitat, simulated environments are prone to error. Should the captured Bigfoot detect those differences, altered behavior patterns may emerge. Additionally, with only one live Bigfoot to study, researchers would gain no knowledge of its social habits. Finally, one must accept the fact that the actions of the captured Bigfoot only reflect the characteristics of the individual specimen and may not necessarily reflect those of the entire species.

Should you find yourself going Bigfoot hunting, remember always to exercise caution whenever approaching a fallen creature. Although the beast may be lying upon the ground, it may not be completely tranquilized or dead. After allowing a sufficient time to pass, you should restrain the creature so that it does not awaken unexpectedly and go on a wild rampage.

REFERENCES

1. Wanted: Dead or Alive -

   http://www.rfthomas.clara.net/papers/ft93.html

2. Hunters and Bigfoot -

   http://www.zerotime.com/2007/10/04/hunters-and-bigfoot/

3. Professor Says Kill a Bigfoot to Prove It's Real -

   http://www.anomalies.net/archive/cni-news/CNI.0457.html

4. To Shoot or Not to Shoot -

   http://www.sasquatchonline.com/content/view/31/29/

5. Mike Conley's Tales of the Weird: Man Didn't Shoot
   Sasquatch -

   http://www.mcdowellnews.com/servlet/Satellite?pagename
   =MMN/MGArticle/MMN_BasicArticle&c=MGArticle&ci
   d=1173355724283

6. How to Observe Bigfoot and Report the Bigfoot Sighting
   http://www.wikihow.com/Observe-Bigfoot-and-Report-the-
   Bigfoot-Sighting

Also available from the author:

INVISIBLE HUMANS

HISTORY, THEORY, AND APPLICATION

Throughout the ages, man has sought to harness the power of invisibility. An invisible person can enter a room undetected, be ignored by hostiles, gather information from remote locations, or just avoid an unwanted social interaction. The techniques described within this work cover the principles of invisibility based on physical, psychological, hypnotic, yogic, and occult practices. These highly guarded secrets, which have been passed down through the centuries, are finally revealed.

ISBN: 1-4392-0953-7

www.ingramcontent.com/pod-product-compliance
Lightning Source LLC
Chambersburg PA
CBHW051418170526
45165CB00004BA/1873